We Are the STEM Stars

Vol 2:
A Need to Survive Treasure Hunt

Written By: Demetria Hoskins
Illustrated By: Alpha "A" Frierson

Dedicated with Love and Gratitude

If you purchased volume one, you will notice that there is a new character in volume two. Introducing....

Ms. Bee, the STEM teacher!

I dedicate this book to my first and most favorite teacher, my mom, Blondine Hoskins. She helped me love to learn and taught me to never be afraid to try something new or something that might seem difficult.

My mom is known for her infamous pound cake and to always "bee" sweet. See if you can find any bees or pound cake in this book while you are STEM-ming.

Thank you, mom, for all that you have poured into me and still pour into me. I am a success because of you.

I love you. 4ever.

"Only humans need water," someone in the lunchroom shouted.

Echo exclaimed, "That's not true!"

"We should go on a "Need to Survive" treasure hunt," suggested Tek.

"Huh? WHAT is a "Need to Survive" treasure hunt, Tek?" asked Engie.

"Well, we get our science journals, look for things that plants and animals need to survive, and record our findings," explained Tek.

This will be fun. Let's go!

"Our classroom fish definitely drink water in their aquarium," Tek observed.

"Ms. Bee makes sure that we always water our plants by the window," said Echo.

"Yes, plants need water to grow and plants get vitamin D from the sunlight," Engie stated.

“Ms. Bee, can we go on a “Need to Survive” treasure hunt?” Tek asked.

"That would be a great way to start off our science for today! Line up," said Ms. Bee.

Engie - "Humans drink water all the time. Water is good for us."

Matt - "I wonder what we will find outside."

"I see a bird drinking water from that puddle. Write it in your journals," said Echo.

"Look at that cat just basking in the sun," giggled Engie.

"Yes, that cat is lying down getting some good old Vitamin D from the sun," explained Ms. Bee.

"We are, too. Humans get vitamin D from the sun," said Matt.

Engie yelled, "Oh no, it is starting to rain!"

"That dog, across the street, ran into its doghouse," said Tek.

"Let's hurry inside and record our data on a T-chart," said Ms. Bee.

T-Chart

Plant/Animal	Need
fish	water
plants	water
humans	water
bird	water
cat	sun
humans	sun
dog	shelter
humans	shelter

(You can add more.)

"Do you see any patterns or similarities from your observations?" asked Ms. Bee.

“Yes, we do!”

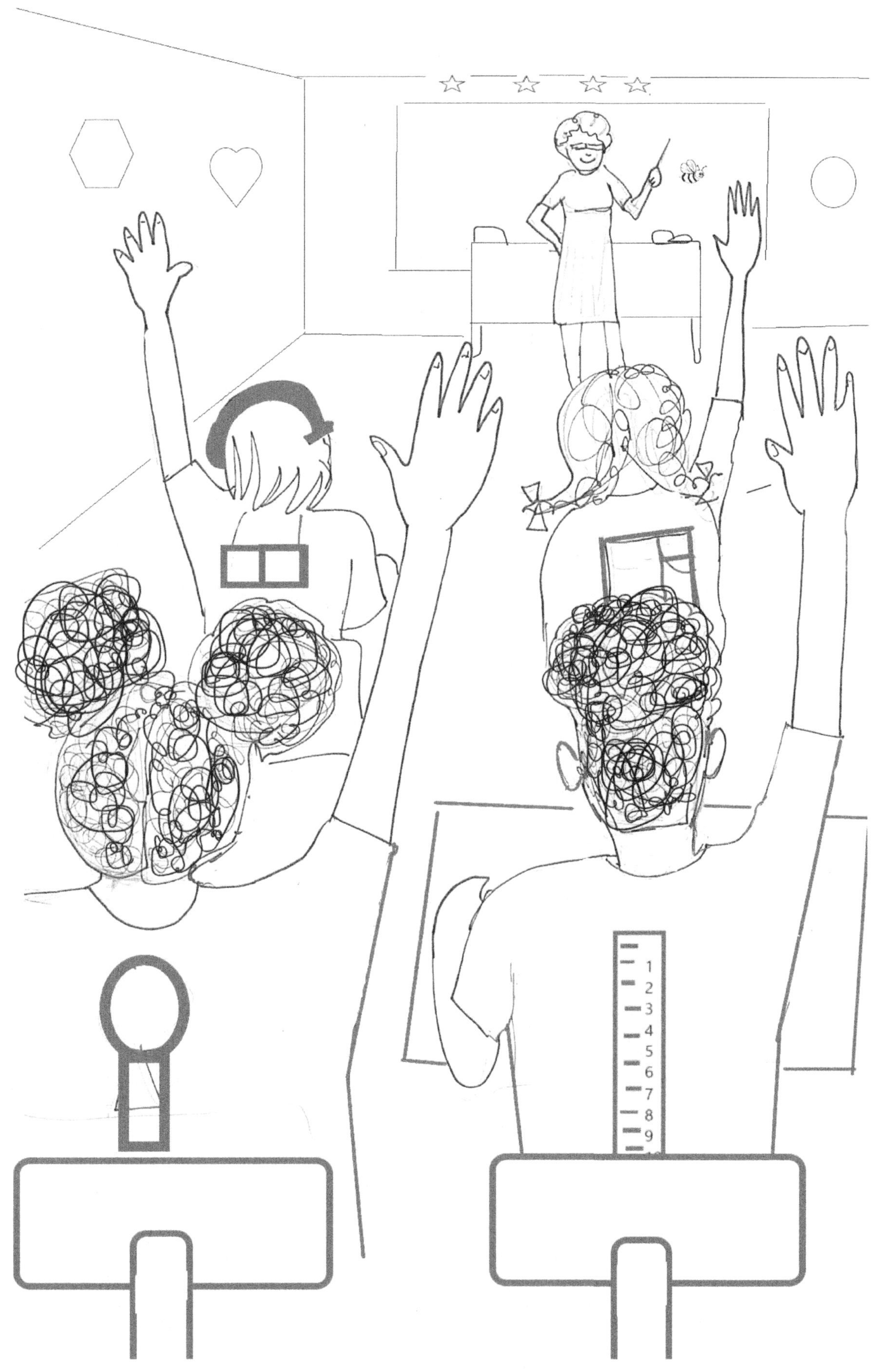

"Excellent! Copy the chart in your science journals. We will add more data to our chart, make comparisons, and find patterns tomorrow," directed Ms. Bee.

"Ut-oh, here comes the homework," whispered Echo.

"Ok class, be prepared to share 1 survival need for a plant or an animal and where you found your information," said Ms. Bee.

Lion

Engie presents:

"Squirrels need 2-3 tablespoons of water per day to survive. My information was from a magazine about animals."

Tek presents:

“A tree needs about 11 gallons of water per week to survive. I found my information on the internet.”

Matt presents:

"A lion is a carnivore and eats 11 to 15 pounds of meat per day. I had a book about lions in my room."

Echo presents:

"An elephant is an herbivore and eats 330-375 pounds of plants per day. I watched a documentary on TV."

1
2
3
4
5
6
7
8
9
10

Squirrel and Tree Water Comparison

"I'm crunching numbers here, guys. So, a squirrel needs 2-3 tablespoons of water per day which equals about 1 ½ cups of water per week, right?" asked Matt.

"Right!" exclaimed Echo, Engie and Tek.

"But a tree needs 11 gallons of water per week to survive and 11 gallons is 176 cups," said Matt.

"The tree, obviously, needs a lot more water per week than the squirrel but they both need the water to survive," Tek explained.

Water Needed for a Squirrel

FREQUENCY NEEDED	AMOUNT OF WATER NEEDED
1 DAY	2-3 TABLESPOONS
1 WEEK	1 ½ CUPS

Water Needed for a Tree

FREQUENCY NEEDED	AMOUNT OF WATER NEEDED
1 DAY	1 ½ GALLONS = 24 CUPS
1 WEEK	11 GALLONS = 176 CUPS

“I’m passing out a chart of needs for you all to complete. Work with a partner and turn it in at the end of class,” said Ms. Bee.

Needs to Survive Chart

	water	food	shelter	sun
fish				
plant				
human				
bird				
cat				
dog				
squirrel				
tree				
lion				
elephant				

Complete this chart. Put a checkmark under the categories that animals, plants and humans need. Have an adult check your data.

Engie said, "Mrs. Bee, I see patterns that animals and plants have in common when it comes to their survival needs."

"Do you all see any survival need similarities or patterns from our Need to Survive treasure hunt?" asked Ms. Bee.

"Yes, we do!" exclaimed the STEM Stars.

“What patterns or similarities did you find between plants’ and animals’ survival needs?”

You, too, can be a STEM Star!!!

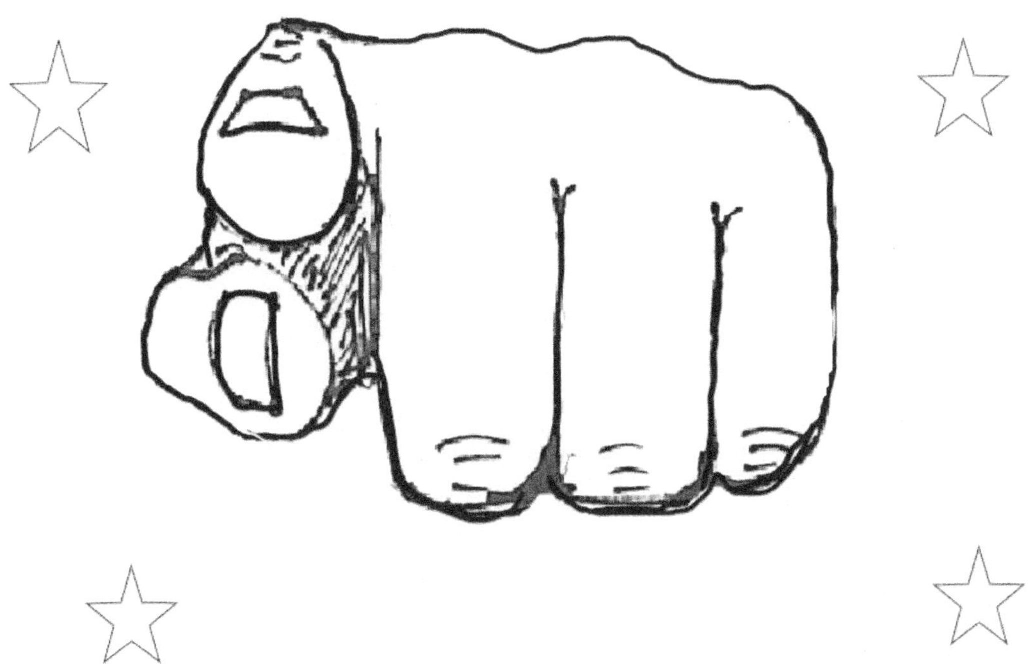

See you next time!!!

Vocabulary Word/STEM Terms

Need: a thing that is required or essential

Survive: continue to live or exist

Shelter: a place giving protection from bad weather or danger

Herbivore: an animal that feeds on plants

Carnivore: an animal that feeds on flesh

Photosynthesis: the process by which plants use sunlight, water, and carbon dioxide to create oxygen and energy in the form of sugar (also referred to as Light Energy)

Similarities: having a likeness or resemblance

Pattern: regular and intelligible forms or sequences that can be found throughout nature

T-chart: Graphic organizer, shaped like a "T," used to separate information into two categories

Gallon: a unit of liquid capacity equal to 128 ounces or 16 cups

Cup: a unit of liquid capacity equal to 8 ounces or 16 tablespoons

Tablespoon: a unit of liquid capacity equal to ½ an ounce

Made in the USA
Middletown, DE
27 February 2023